I0224603

I GROW CHRONIC II

MASTERS OF THE CLOSED SYSTEM

BY MR GREEN

IGROWCHRONIC.COM

I GROW CHRONIC II: MASTERS OF THE CLOSED SYSTEM. *This grow guide compares and contrasts three indoor grow rooms filled with pounds of highest quality cannabis flowers. By JD Green*

Published by:
Mr. Green's Chronic Entertainment Company.
www.igrowchronic.com
Los Angeles, California.

ISBN- : 978-0-578-72686-1

COPYRIGHT © 2021 JD Green, Mr. Green's Chronic Entertainment Company.
PHOTOGRAPHS © 2021 JD Green, Mr. Green's Chronic Entertainment Company.
VIDEO © 2021 JD Green, Mr. Green's Chronic Entertainment Company.

NOTICE OF RIGHTS
All rights reserved. No part of this book may be reproduced or transmitted in any form by any means, electronic, mechanical, photocopying, recording or otherwise, without the prior written permission of the publisher. For more information on obtaining permission for reprints and excerpts goto the contact page www.igrowchronic.com.

NOTICE OF LIABILITY
The information in this book is distributed on an "as is" basis, without warranty. While every precaution has been taken in the preparation of the book, neither the author nor the publisher shall have any liability to any person or entity with respect to any loss or damage caused or alleged to be caused directly or indirectly by the instructions contained in this book.

TRADEMARKS
All product names identified throughtout this book are used in editorial fashion only and for the benefit of such companies with no intention of infringement of the trademark. No such use, or the use of any trade name, is intended to convey endorsement or other affiliation with the author of this book.

DISCLAIMER
This book does not promote the breaking of any laws regarding the growing of cannabis. This book is an exercise of the Authors Freedom of Speech protected by the First Amendment of the United States Constitution and by Common Law and is meant only to educate and entertain and never intented to cause damage or harm.

I GROW CHRONIC II

MASTERS OF THE CLOSED SYSTEM

BY MR GREEN

TABLE OF CONTENTS

INTRODUCTION

This is an intermediate grow guide book intended to take you and your grow to the next level up in cannabis cultivation.

Mr. Green tours three productive, awe inspiring indoor grows that you can manage. All three rooms use approximately the same wattage in lights and have similar square footage to work with, so you can really compare the different grow techniques and equipment used, so that you can decide what will work best in your grow room.

WELCOME BACK!

and welcome to room number

ONE.

You may want to take notice
as this grow room is

THE GOOD ROOM.

ROOM ONE. This is a really nice grow room, better than most. The caliber of the cannabis flowers grown in this space is of the utmost quality. The crystals are there, the herb is sticky-gooey and delicious but something is missing. The weight. Nobody will notice this, except the grower. He knows this because he hears stories of "other growers", with similiar set-ups, pulling more weight. As a grower, it makes you wonder if you are getting as much flower as possible out of each harvest or if you are leaving weight on the table. Here, we begin our tour of three solid grow rooms. Note the similarities of all three spaces. Each room has many things to learn from as they all throw fire. See pictures to the right. > > > > > > > > > > > > >

This room is the one with the greatest challenge as it is not operating up to its full potential. This room is not running as a closed system and therefore is leaving 25% of its potential flowers unrealizied.

LIGHTS: 2x's 1000watt HPS + 2x's 600watt HPS. Total 3200 watts of artificial light.
Four SunSystem™ 6 inch air cooled lamp hoods with glass.

WATERING SYSTEM: Air-O-Ponic Hydroponic system. The General Hydroponics AeroFlo 60™ Hydroponic air hose feeding system.

AIR ENVIRONMENTIAL CONTROLS: One large 12 inch exit fan with carbon filter, passive intake vent with carbon filter, several room fans, one 8 inch carbon filter scrubber fan, a dehumidifier and a CO2 generator.

ROOM DIMENSIONS: 8 feet wide by 12 feet long with just over 7 feet of height. Approximately 100 sq. feet. Low but manageable ceilings.

GENETICS: Pineapple Express; Super Kush.

FLOWERS PRODUCED: 3.25 pounds.

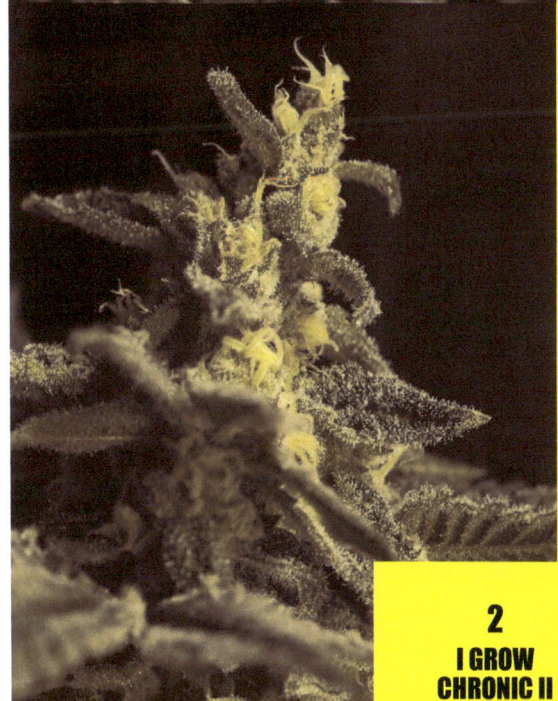

Lights. If you are running a room without an air conditioner, like this room, you will need to vent and air cool your HPS lights. Room one has 4 lights with glass sealed hoods, which are all connected to each other with 6 inch air-vent tubing. The last hood is then connected to a fan with a 6 inch vent tube which pulls the hot air produced from the lights out of the grow room. These types of hoods have been in the grow industry for 30 plus years and are still used today. It's the light bulbs and ballasts that are constantly changing and updating. Growers naturally gravitate towards, once proven, newer technology as we always want more, better, faster results. Everything else in this room, concerning the lights, is straight forward and basic.

The room has two 600 watt HPS lamps and two 1000 watt HPS lights with electric ballasts producing a total of 3200 watts. It is actually a nice set-up with the light hitting every part of the canopy. There is a heavy duty pool timer used as the controller and that's all there is to running the lights in this room.

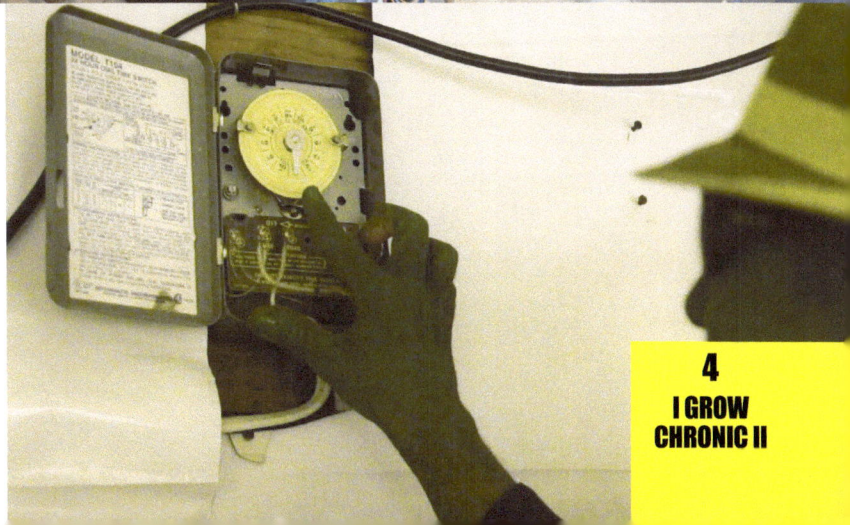

Watering System. The General Hydroponics AeroFlo 60™ unit has been around the scene for over 30 years! This legendary system speaks for itself as a leader in growth and yield rates. It is a self contained unit with a re-circulating reservoir of nutrient rich, oxygen infused water that sprays directly onto the roots of the plants. There is no soil to move, as the plant roots grow through 3 inch plastic cages filled with hydroponic pebbles and into the square white chambers. The key to success when using a system with a recirculating resevoir, such as the AeroFlo 60™, is to use a water chiller. It is very important to cool the nutrient water that feeds the plants to 68° F. High nutrient water temprature, above 72° F, can induce root rot. Use a water chiller to keep it cool.

Remember as well, the qualtity of the water matters. This grower does not use Reverse Osmosis water but rather filters the water using a Aquasana™ whole home residential filtering system with pre and post filters. This carbon filter system lowers the PPM (parts per million) of the water to between 100 and 300 which is good for growing plants. The grow space also has hot and cold running water with a drain in the floor which makes for better work flow and clean-up.

Air Environmental Controls. The airflow in room one is very simple. It has one 12 inch fan that pulls the air out of the room through a large carbon filter. This air pulling action creates a vaccum effect in the grow space which then pulls fresh air into the grow room from the outside. The carbon filters are a necessity in preventing strong chronic smells from invading the neighborhood. The air intake is passive (no fan) and is connected to a carbon filter to prevent bugs and smells from traveling into or out of the grow room. The grower also utilizies a scrubber which is basically a vent fan set on top of a carbon filter. What this does is clean (scrub) the air in a given room by recirculating the air through the carbon filter. No vent tube is needed as the scrubber system's job is to clean the air in a room, not deliver that air to the outside.

The room also has several fans blowing air around, a heater for cold nights and a de-humidifier in case it gets too humid in the space.

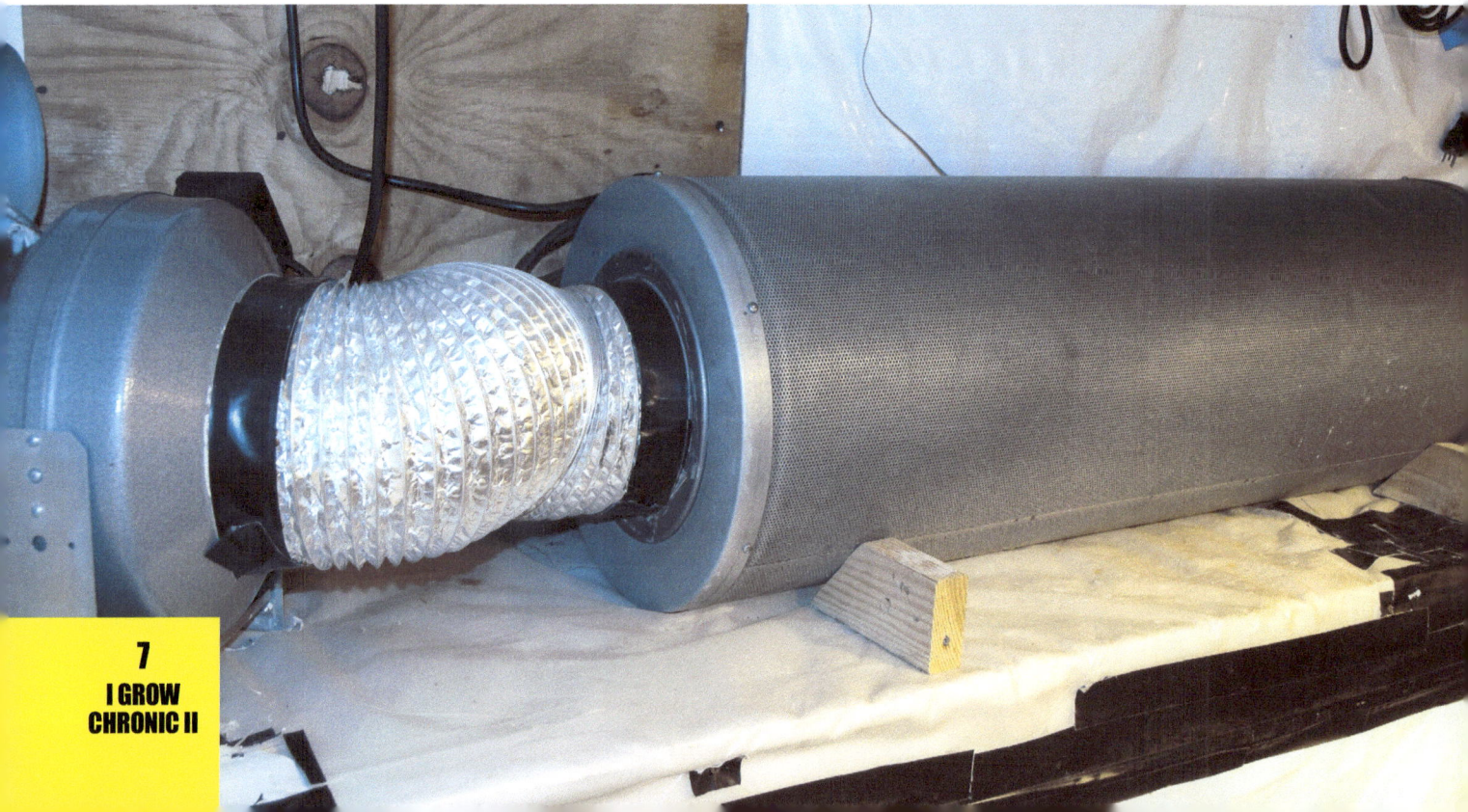

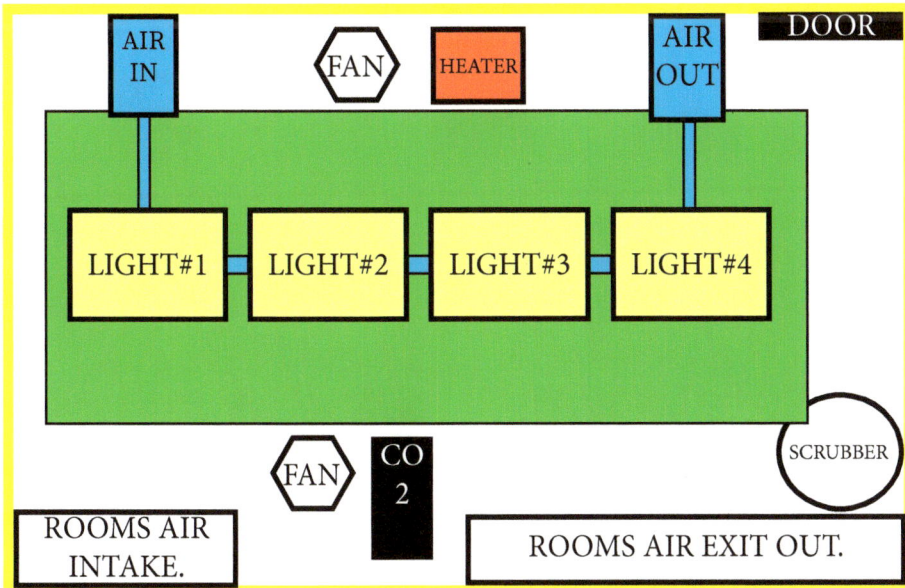

					DOOR
AIR IN	FAN	HEATER		AIR OUT	

LIGHT#1   LIGHT#2   LIGHT#3   LIGHT#4

FAN   CO2   SCRUBBER

ROOMS AIR INTAKE.

ROOMS AIR EXIT OUT.

CO2. Carbon dioxide. Grower number one knows the importance of this gas as it relates to plant life. One of the most amazing scientific phenomenons on planet Earth takes place in your grow room, the converting of carbon dioxide into oxygen.

Plants breathe in CO2 and exhale O2. We humans exhale CO2 and inhale O2. It is no wonder that we get along so well together. When plants breath in CO2 it gives them the energy to photosynthesize light into sugars that provide the plant with energy to live and grow. There is ample amounts of CO2 in the air we breathe for plants to live but as a cannabis farmer you want your plants to not only live but to thrive. The cannabis industry, through years of trial and error, has discovered that there are gains in production to be made with CO2 enrichment. 25% more flower is possible when growers use a carbon dioxide system to immerse the garden in this gas. That sounds easy enough... we just set it up, turn it on and go!

NOT SO FAST!

There is more you need to know before you grow...

Now we really do have a PROBLEM

and not an easy solution. The grower in room one is not getting the 25% increase in yields even though he is blasting his room with CO2. At this point it seems like he is just wasting money on propane to run his burner. This is happening because his exhaust vent fan sucks all the CO2 out of the room before the plants have a chance to utilize the gas. So in an attempt to prevent this situation from occurring he turns OFF the exhaust fan. This makes it all good in the grow room for about five minutes at which time the overall room temperature starts to raise rapidly. This gets dangerous as high temperatures kill plants. So to prevent the rooms temperature from climbing any higher, the exhaust vent fan must turn back ON and in that process it clears all the enriched CO2 air out of the room before the plants truly have a chance to utilize it.

That's how this room runs. Inefficiently.

There is a constant battle in this space between the temperature and the CO2 levels. It is unable to maintain consistent CO2 levels so that the garden can benefit from the gas. When CO2 enhancement is run correctly it can deliver a 25% increase (or more) in production yields but the way this room is running, the plants are maybe getting a 5% to 10% increase, if that.

Let me take you to the next level of indoor cannabis production that solves this very problem. I will show you the inner workings of the CLOSED SYSTEM.

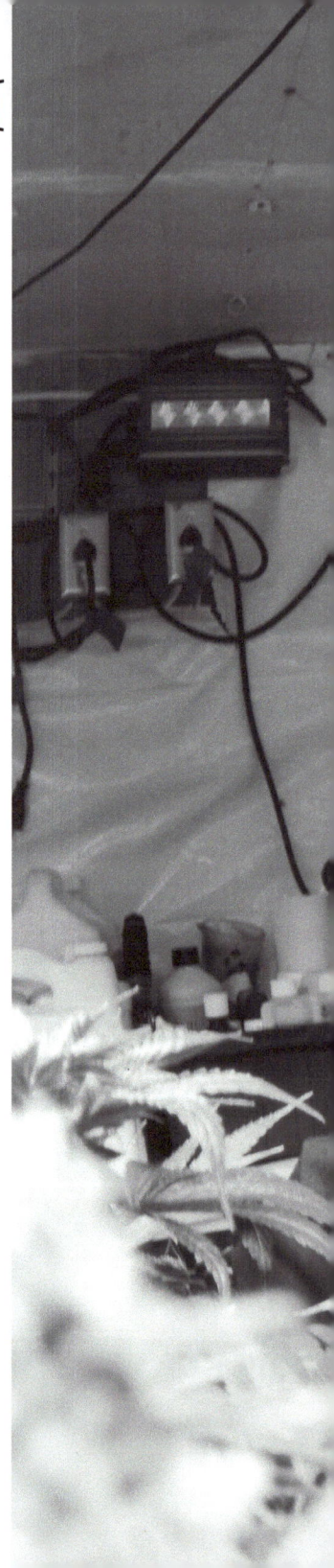

Hello Friends! Come inside room number two and close the door.

WELCOME TO THE

BETTER

GROW ROOM.

the next level of grow room perfection...

THE CLOSED SYSTEM.

ROOM TWO. This room is better! Like room one, the caliber of the cannabis flowers grown in this space is also of the utmost quality but this room produces *25% MORE FLOWERS!*

AIR ENVIRONMENTAL CONTROLS: Closed System:

Fujitsu™ 30,800 BTU mini-spilt air conditioner unit; CO2 generating system; computer with sensors; multiple room fans & de-humidifier.

LIGHTS: 2x's 600 watt HPS & 2x's 1000 watt HPS systems totaling 3200 watts.

WATERING SYSTEM: Hand water nutrient feed into COCO medium.

ROOM DIMENSIONS: 10 feet x 10 feet with 8 foot ceiling. 100 square feet.

GENETICS: Pineapple Express, Super Kush and a few Blue Dreams.

FLOWERS PRODUCED: 4.5 pounds. A solid 25% increase over room one. using the same wattage (lights) and genetics.

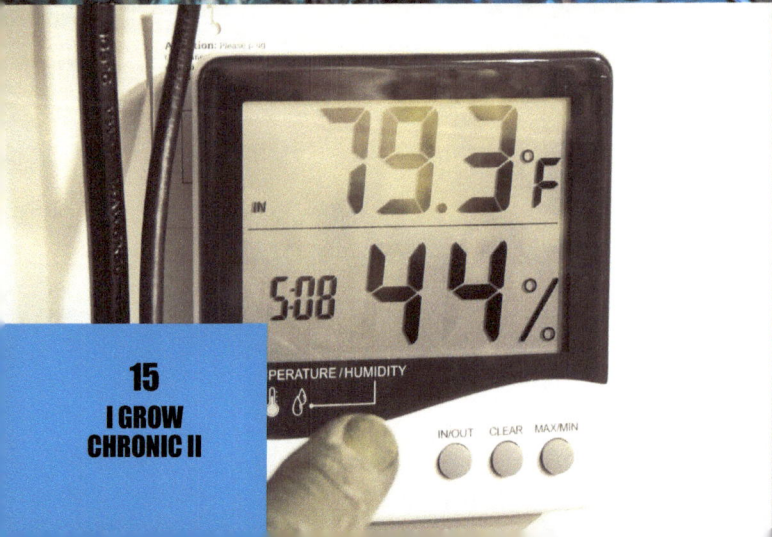

79.3°F

IN

5:08 **44%**

TEMPERATURE / HUMIDITY

IN/OUT   CLEAR   MAX/MIN

15
I GROW
CHRONIC II

AIR ENVIRONMENT. When thinking about a closed system I want you to think biosphere. "The Biosphere is the worldwide sum of all ecosystems on Earth. It can also be defined as a self-regulating closed system" -Wikipedia.com

That's what we are interested in... a self contained, self regulating, closed ecosystem. That's what we want to build, an artificial biosphere in your grow room. That is exactly what grower number two has accomplished with this closed system. He has created an environmental vacuum in which life can thrive for plants. This sounds difficult and excessively scientific but it is not that much more challenging than building a regular grow room. The closed system grow room uses more equipment and requires a little more knowledge and attention but the results speak for themselves.

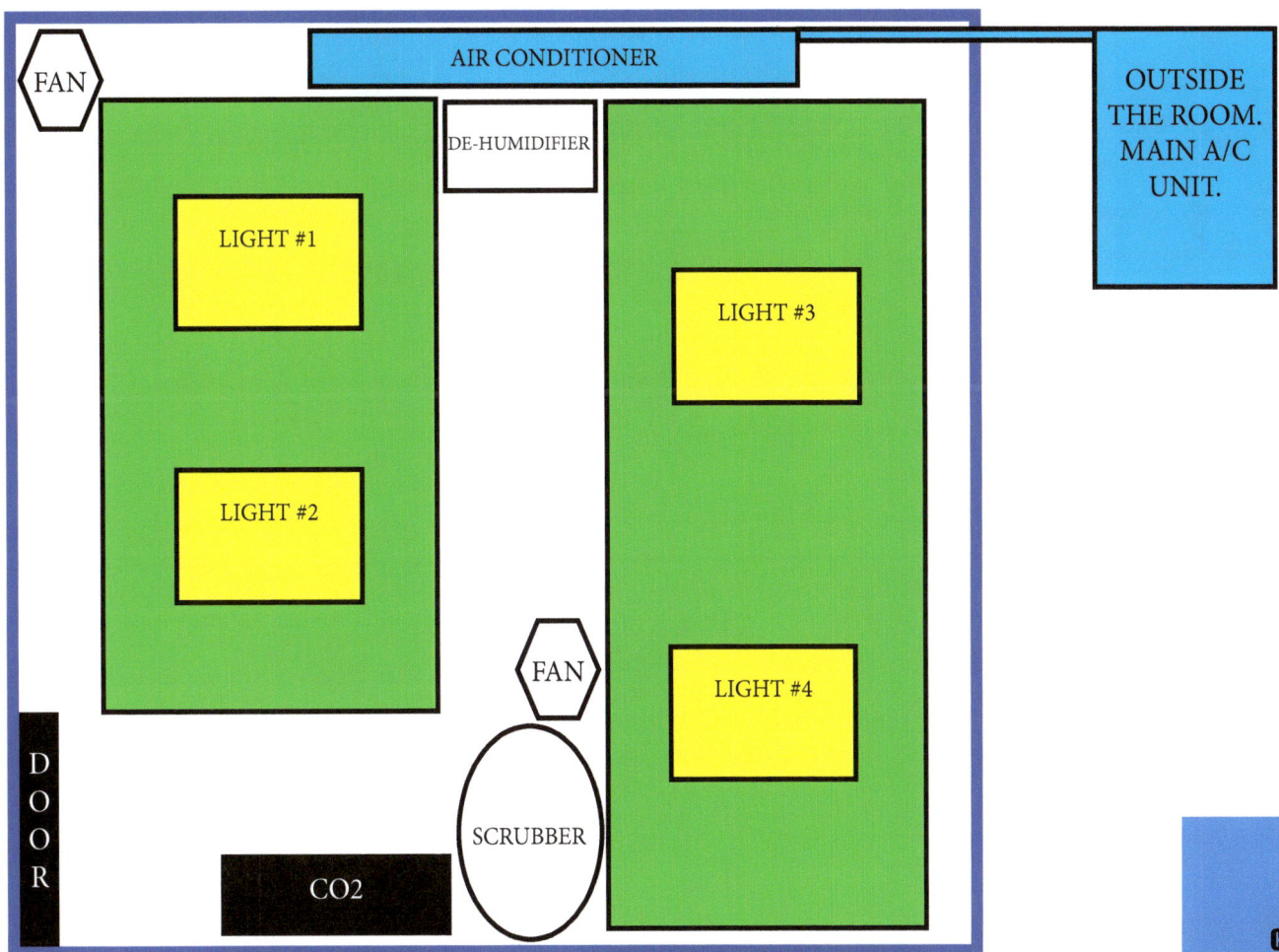

FAN

AIR CONDITIONER

OUTSIDE THE ROOM. MAIN A/C UNIT.

DE-HUMIDIFIER

LIGHT #1

LIGHT #3

LIGHT #2

LIGHT #4

FAN

DOOR

SCRUBBER

CO2

AIR CONDITIONER. Lights make grow rooms hot. Any temperature above 90° F is dangerous and can get deadly for plants. This is the reason air-cooled lights, like the ones used in room one, were invented and became so popular with growers back in the 1990's and 2000's. You can basically run a grow room's air system with a few vent fans. Alas, problems can arise using this technique because the air coming in to cool down the grow room is from outside and outside air can get hot in the summer time. During warm seasons, this system becomes ineffective in lowering the grow room's temperature. Think about driving around in the car on a 100° F day. If you try rolling down the windows to cool off you will just blast hot air into the car but if you roll the windows up and turn ON the air conditioner, you will cool off. Your car runs like a closed system.

The type of air conditioner you install does matter as you'll want to get a powerful mini-spilt type system. Mini-spilt A/C's work well for grow rooms because the hot parts, the engine of the A/C unit, stay outside of the room and do not affect the temperature of the grow space. This room has one 30,800 BTU mini-spilt air conditioner unit manufactured by Fujitsu™. It does a great job of cooling the room with four lights.

The A/C grow rule goes:
Minimum 4,000 BTU for every 1000 watts of light. When adding other grow equipment to the equation it is safe to have up to 7,000 BTU for every light.

What temperature is best for growing cannabis?
High 70°'s is best for a rooms running without CO_2 enhancement as higher temperatures can stress the plants and slow their growth rates. Growers that operate a CO_2 system can run their rooms hotter. With elevated CO_2 levels, plants can withstand overall higher temperatures, as most closed system growers run their rooms in the low to mid 80's.

Now that we have had an air conditioner information download, we are on the right track to success, but there is still some more critical information to understand as we move forward.

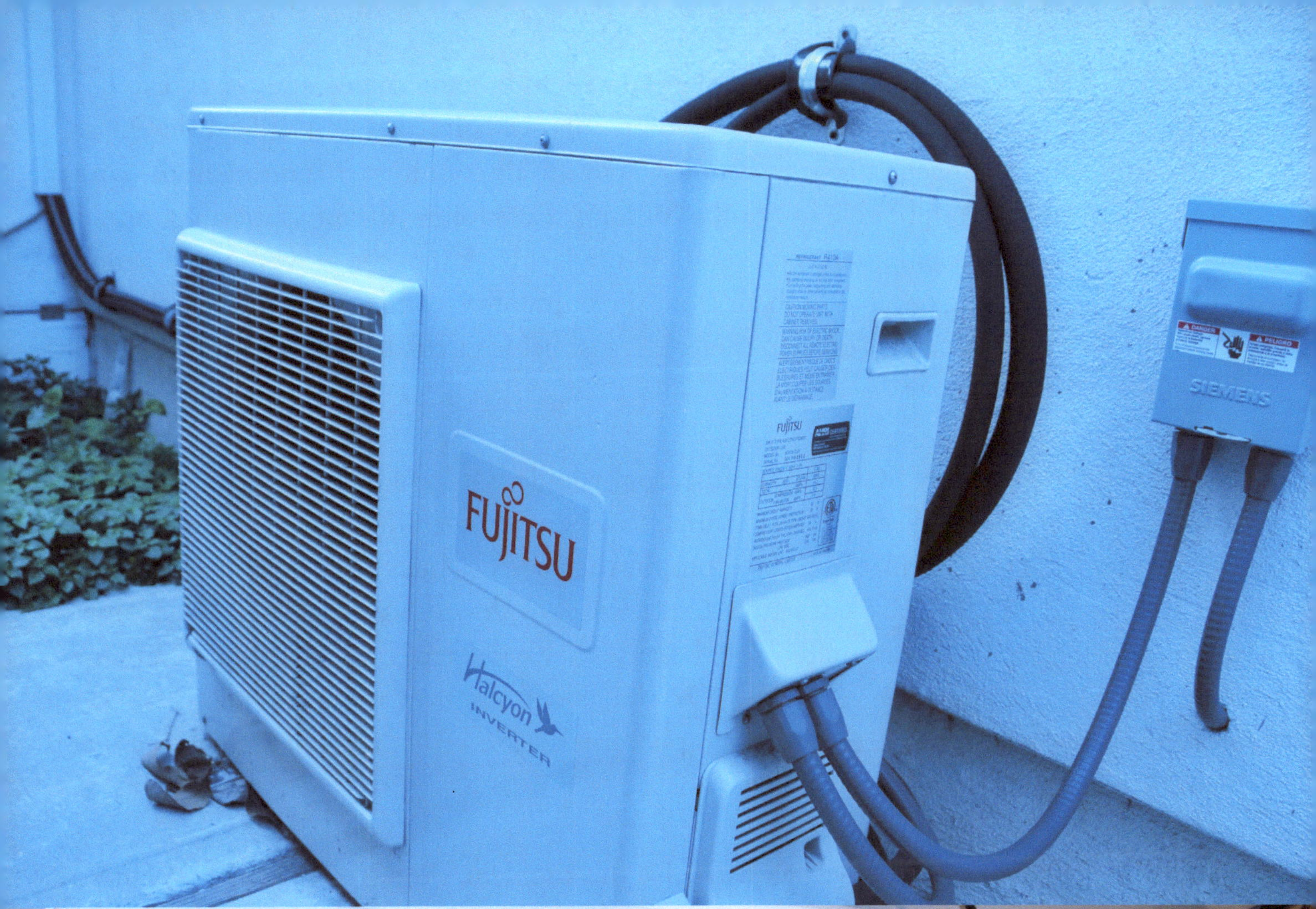
FUJITSU
Halcyon
INVERTER
SIEMENS

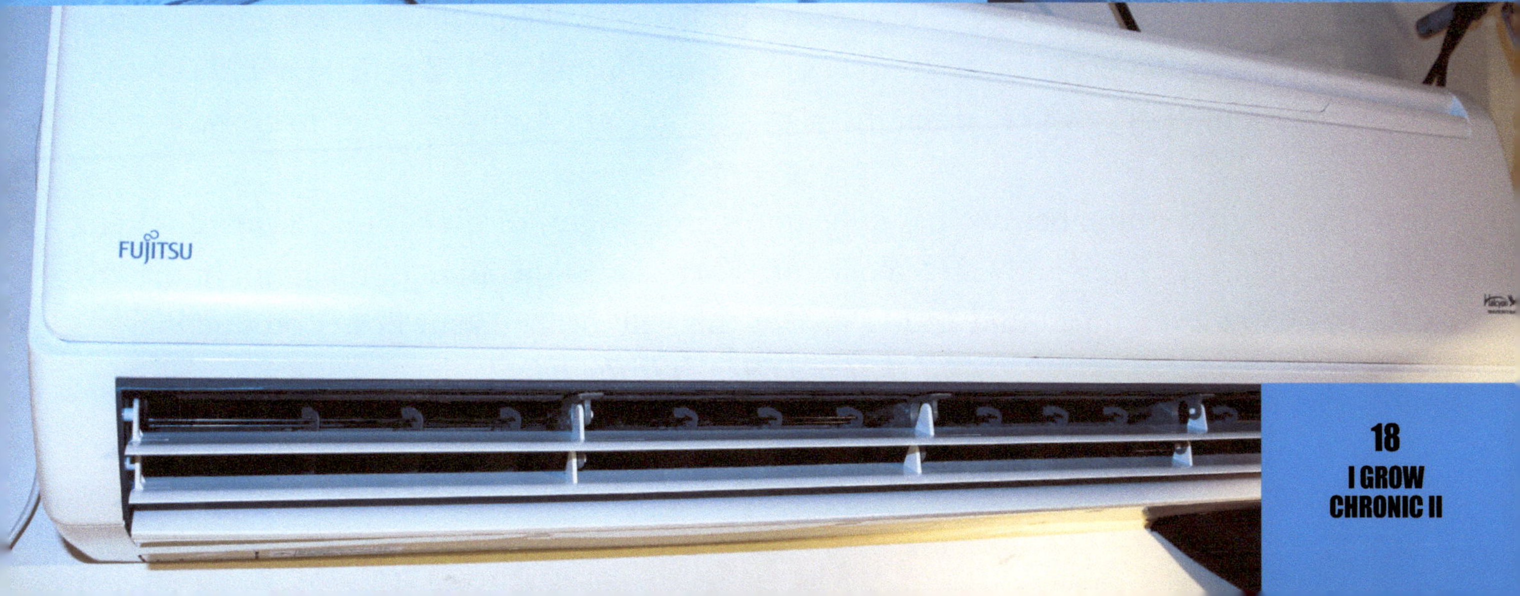
FUJITSU

CO2 SYSTEM. At this point, the air in the grow room is being cooled and that is excellent but now we have another big challenge. If you recall, plants inhale CO_2 and exhale O_2, so if there is no air exchange in the grow room, due to the air conditioner running in the closed room, the plants will not be able to breathe for long. Let's follow the air as the air conditioner pulls it out of a grow room, then cools it and re-sends it back into the grow room. There is NO air exchange going on. No fresh air from the outdoors is coming into the grow room. It is the same air that has always been in there, it is just being circulated and cooled through the A/C unit.

Remember, the plants need CO_2 to breathe, so if you do not have fresh air coming into your grow room your plants will continue to inhale CO_2 and exhale O_2 until the point at which there is very little CO_2 available in the room for the plants to breathe. The room will be saturated in oxygen (O_2). The plant's growth rate will slow down to a stand-still. The plants will be starving for CO_2 (carbon dioxide) because they will have converted all the CO_2 in the room into O_2. There will be very little CO_2 available in the grow room and plants will stop growing as they will not have the energy to photosynthesize. Even with all the lights and all that nutrient water, the garden will fail.

The garden will not succeed unless we can solve this problem. We need to somehow convert the O_2 in the room to CO_2. We are already familiar with the piece of grow equipment that does just that, the CO_2 generator (burner). We could install a CO_2 generator and convert the O_2 in the room back to CO_2. Then the plants can breathe again and start to grow.

That is what grower number two has done to his grow room, turning it into a closed system. Get yourself an A/C unit and CO_2 generator. Keep the temperature of the room in the low 80°'s and CO_2 levels up around 1300 PPM, as this will increase your flower production.

And that my friends is the trick!

THE TRICK TO 25% GAINS.

COMPUTERS AND THE AIR ENVIRONMENT. It sounds unbelievable that running an A/C unit that cools off the grow room a little and then saturating the room with CO2 gas would put the plants in such a good atmosphere that they'll grow 25% more flowers, but it is true. I'm not trying to oversimplify this process as it would be impossible to run a CLOSED SYSTEM without the use of modern day technology.

There are several CO2 computer controllers on the market to choose from and grower two is using the Sentinel - Advanced CO2 PPM Controller™ with great success. The main job of this computer unit is to monitor the CO2 (carbon dioxide) levels and control the activity of the CO2 generator in the grow room. The computer sensors are used to measure the PPM (parts per million) levels of CO2 gas in the grow room. Plants can utilize up to 1500 PPM of CO2 in the grow room. Anything over 1500 PPM is a waste of fuel, energy and gas as the plants can not utilize these high levels of CO2 gas. Extreme high levels of CO2 gas, above 1500 ppm, can actually harm the plants so be aware and don't damage your garden. Also, as a human, try not to breathe in high levels of CO2 gas when working in the room as we

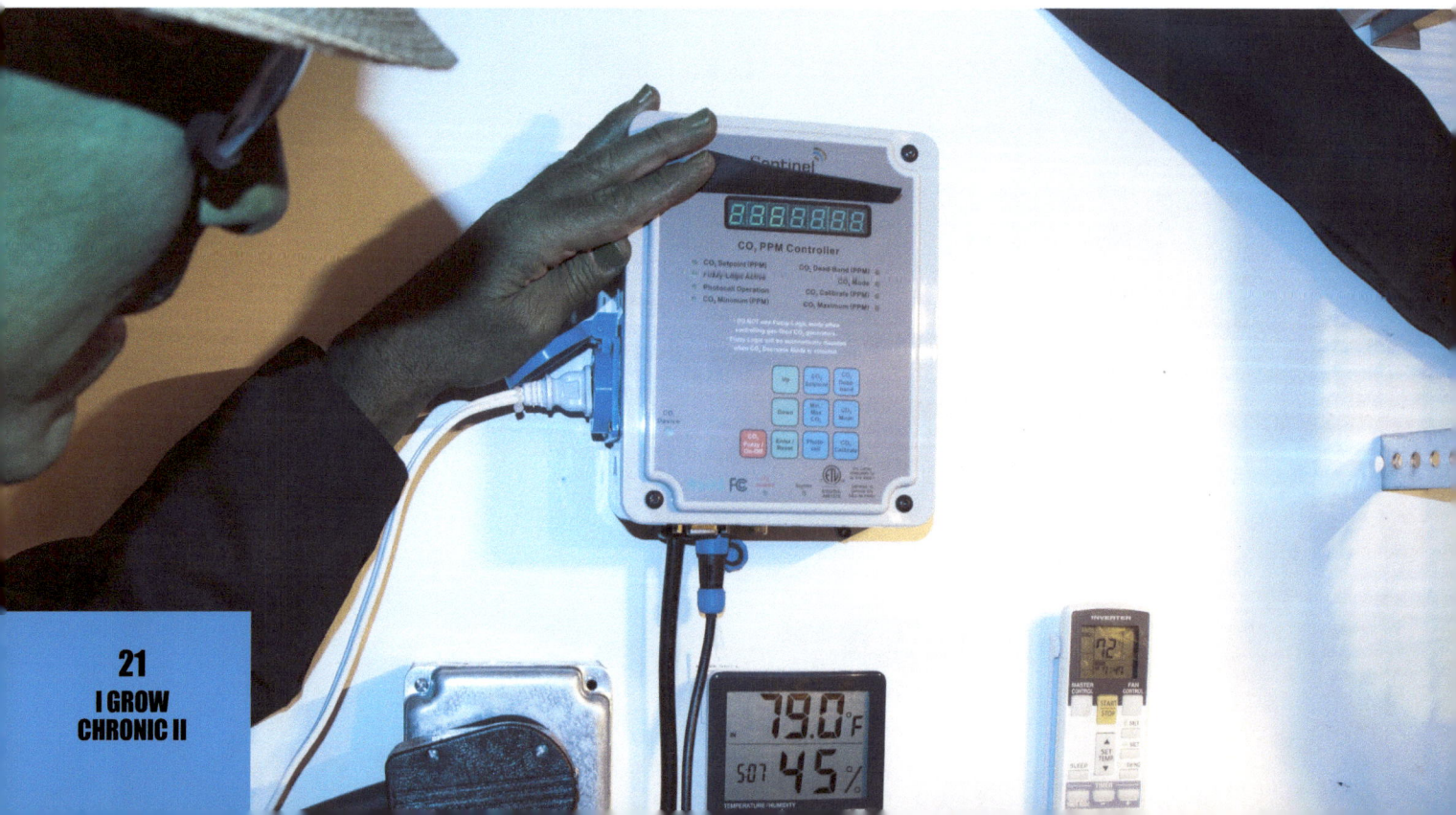

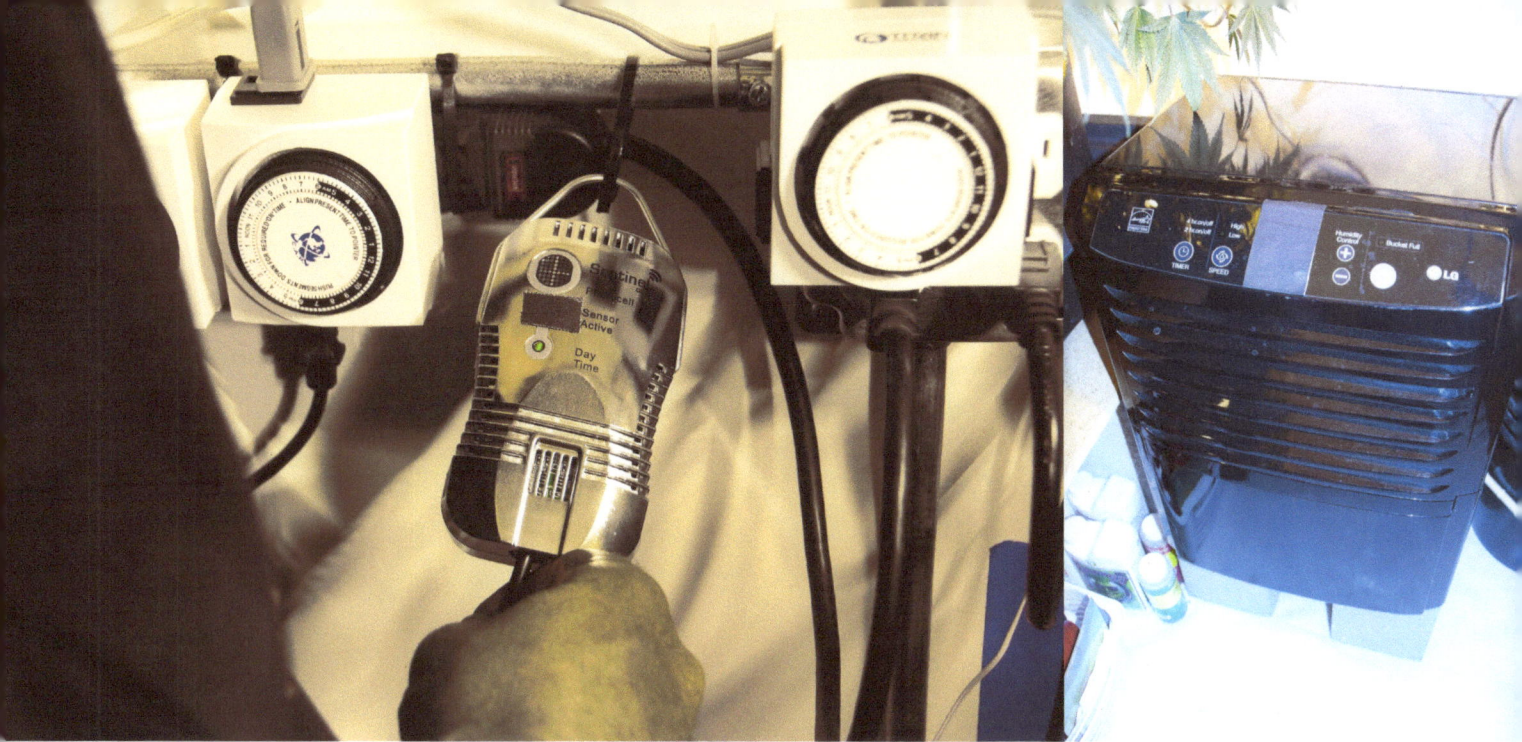

require O2 (oxygen) to live and these elevated levels of CO_2 are unhealthy for us. The computer has been set to raise the CO_2 levels in the room to 1300 PPM with a 100 PPM buffer range. The sensor measures the amout of CO_2 gas in the room and communicates this information to the computer. When the CO_2 levels in the room drop below 1200 PPM the generator (burner) is turned ON. CO_2 gas fills the room until the sensors measure 1300 PPM in the space and the computer turns OFF the gas burner.

Along with the air conditioner unit and the CO_2 generator this room is using many fans to keep the air moving in the room, a de-humidifier to lower the humidity and a hanging scrubber to clean up the chronic, skunk odor in the room.

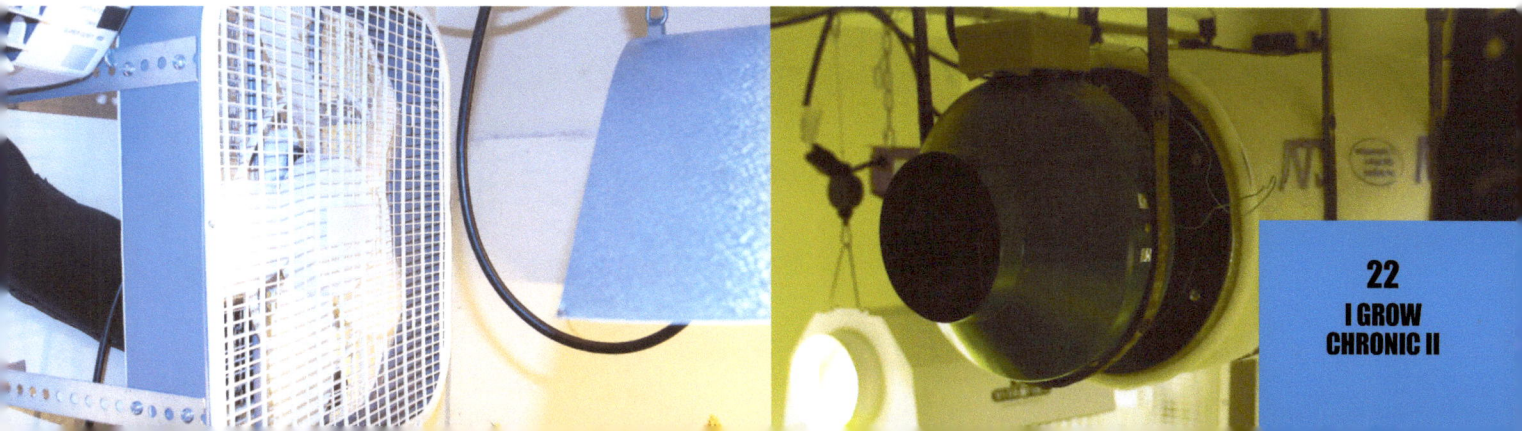

LIGHTS. Total 3200 watts.
2x's 600 watts HPS + 2x's 1000 watts HPS.

LIGHTS. Room two is running the same light wattage as room one and is pulling 25% more flowers, which in this room adds up to 1.25 more pounds. Both rooms are lit up with HPS bulbs and electric ballasts totaling 3200 watts of light power. The only major difference between both rooms is the CLOSED SYSTEM. Note that when using the air conditioner the farmer bypasses the need for an air cooled lighting system. Not only does he not need that extra equipment to cool the lights but he actually gains lumens of light. Even though room two is using the same light wattage, his canopy is recieving up to 20% more light because he is not enclosing the lights in glass. The glass used to seal the light hoods in room one are not necessary in room two, as the powerful A/C unit does an excellent job of cooling the room, lights included. Just removing the glass from the hoods will increase the amount of light available to the plants. More light equals more flower!

Instead of a pool timer, this grower uses a Titan Controls - Helios 12™ light controller which is easy and convenient to set up and use but a little more costly to purchase. So the technique to getting more lumens out of the lighting system is to remove the hood glass, but this is only made possible when running an A/C unit in a Closed System. Also, remember to replace the old HPS bulbs for new ones every year or risk losing weight and quality in your harvest.

Before we go, check out these steel strut, C shaped channels which are typically used for supporting electrical conduit and pipe in the construction industry. The grower has used these struts to hang all of the equipment in this grow room and I think it has worked out well.

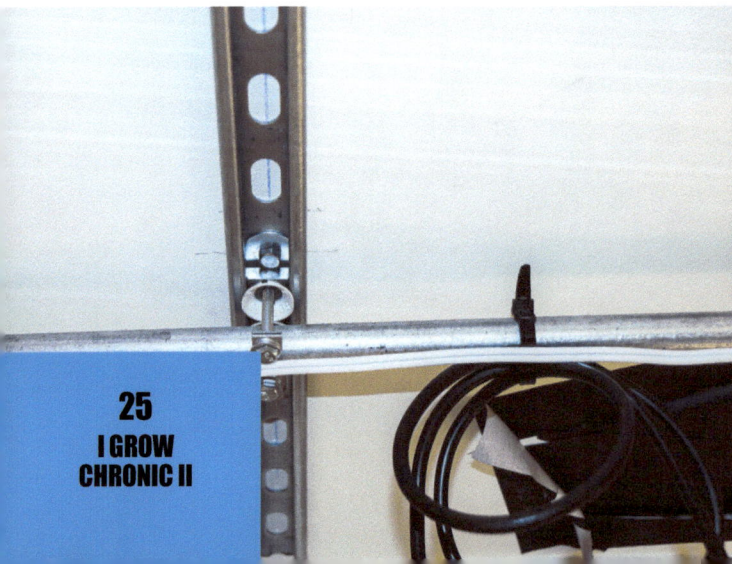

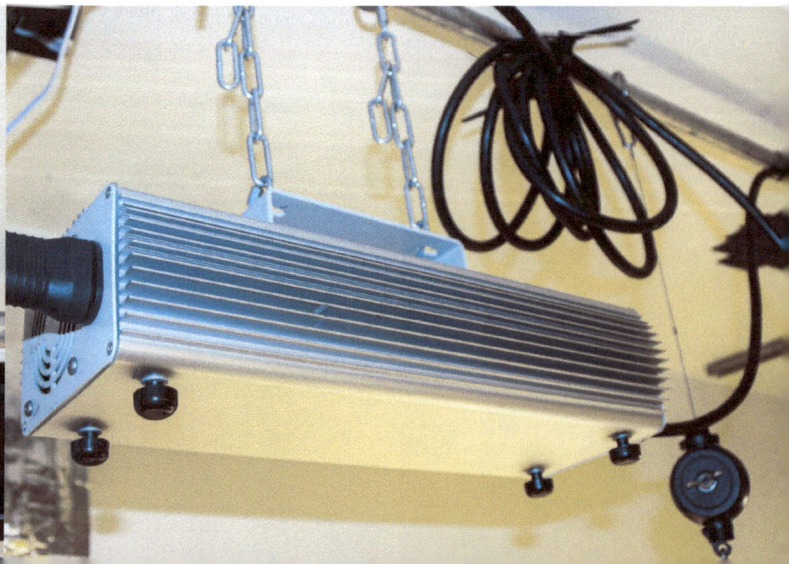

CLOSED SYSTEM growers get the advantage of more light/lumens because they can use open lamp shades instead of glass enclosed hoods that require air cooling. The air conditioner unit does an excellent job of cooling the room, hot lamps included.

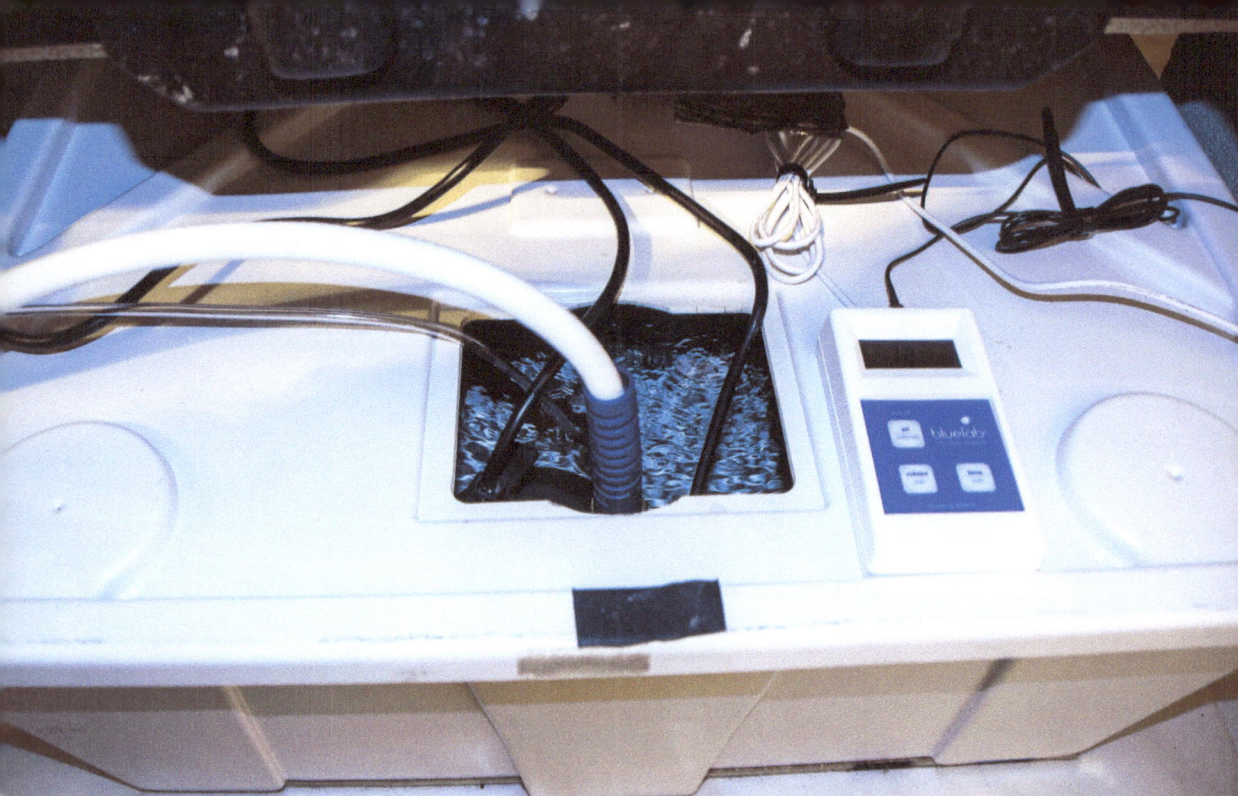

WATERING SYSTEM. Room two feeds the plants using the drain to waste method. He mixes nutrients with reverse osmosis water in a 30 gallon container and hand waters the garden. It takes a few extra minutes every couple of days to do it this way, but this gives him the opportunity to examine each plant and make sure they are all healthy. There is no need for a water chiller to cool the mix as long as it's close to 68° F when feeding. Growers really need to avoid pouring warm water all over the roots, as this can cause root rot and ruin your grow. In a drain to waste system, after watering the plants, the excess water drains off into a collection bucket and is ready for disposal. Some growers complain that running a drain to waste system is too expensive as good nutrients are thrown away. This is actually a valid point but remember the grower has minimal equipment to purchase when using the drain to waste method and the plants seem to really like the freshly mixed nutrient batch.

How do I know this? Because when I get super HIGH, I talk to my plants and they tell me so. It's okay to spend time with your plants, mayby you will start talking to them too?

Remember, you must check the p.H. and the PPM of the water before feeding. A safe range is 6.0 to 6.8 p.H. and 1000 to 1400 PPM of nutrient mix.

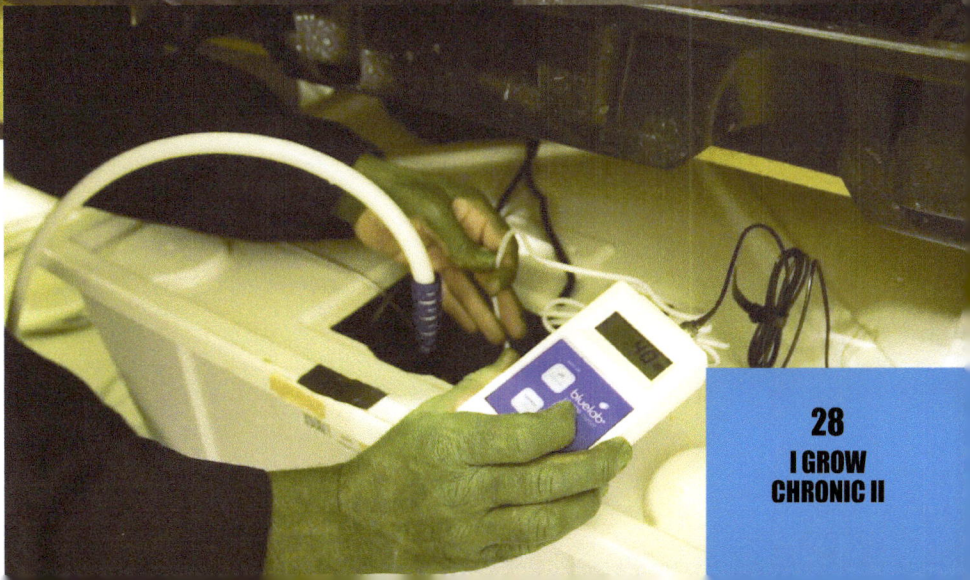

Building a closed system grow room gives us the opportunity to control the air environment so that our plants are thriving. With elevated CO_2 levels at 1300PPM and controlled temperatures in the low 80°'s, the plants can hit their peak production. Room two has pulled it all together and is now getting a 25% increase in yield.

That is amazing but what would you think if I told you that you could do way better than this result? I mean, 25% is amazing but what if you could do better..?

Let us go tour room three where the grower has done just that.
Two and three times that!

Now, this grower is a true master of the closed system as he puts all his knowledge and experience of over 25 years into his grow and it really shows as his garden, of similar size and light wattage, is harvesting over double of what room two is pulling and three times the harvest amount of room one!

BUDDY! What up?

Oh, it's chop chop time!
I'll cruise over to show the crew right now.

BEST

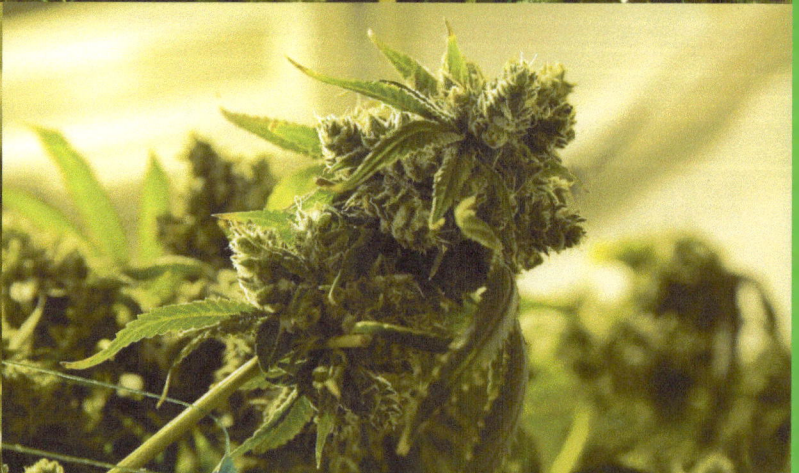

ROOM THREE. This room is the **best room** in our story. The caliber of the cannabis flowers grown in this space is of the utmost quality and the yield is over 3 times that of room one and double that of room two.

LIGHTS: 2x's 1000 watt standard HPS + 2x's 1000 watt *(turned down to 850 watt)* double ended Gavita-Pro™ HPS for a total of 3700 watts of light. 500 watts more.

WATERING SYSTEM: Basic ebb and flow style, spray system with a recirculating nutrient reservoir that pumps the nutrient water over rockwool cubes.

AIR ENVIRONMENTAL CONTROLS: Closed system. Two ton 24,000 BTU mini-spilt air conditioner unit, computer with sensors, humidifer, de-humidifer, fans and scrubber system.

ROOM DIMENSIONS: 10 feet by 12 feet and 8.5 feet high and totals in at 120 square feet.

GENETICS: Gorilla Glue, Herijuana and Pineapple Express.

FLOWERS PRODUCED: Just over 10 pounds!

LIGHTS. This room has two standard 1000 watt HPS lights plus two double ended 1000 watt HPS "Gavita-Pro™" adjustable lights. This configuration can potentially serve up a total of 4000 watts of light, but the grower has the double ended lights turned down to 850 watts for a grand total of 3700 watts of light in room three. The Gavita-Pro™ light wattage is adjustable depending on how much power the grower wants. These double ended lights are actually very powerful and will burn your plants if they are too close to the bulb. With his ceiling just over 8 feet tall, he did not have 3 feet clearance between the bulbs and the tops of the plants, so he decided to adjust the power setting of the lights down to the "low" setting of 850 watts. This room has a 500 watt advantage over rooms one and two, but this 500 watt bump is **not** the sole reason he is pulling double the weight. Of course the extra wattage helps with the overall yield gain, possibly up to a pound can be attributed to the lights, but this guy

is pulling 5 pounds more. Back in the 1990's, a 1000 watt HPS could grow you on average a pound of cannabis and now, thirty years later in the year 2020, a *1000 watt double ended* HPS lamp can grow you one and a half pounds. If the grower is really good, he can harvest just over two pounds per light which is about the maximum I've ever heard of or seen produced in an indoor space.

Again, in a closed system grow room, there is no need for glass enclosed light hoods and air cooled lights as the air conditioner unit cools this room including the lights, efficiently. The digital ballasts are plugged into a "master lighting controller" this one in room three is manufactured by C.A.P.™.

The lighting system in room three is simple and productive. He changes the bulbs out every 6 months so the plants get the most lumens possible. I change mine out every year because I'm a on a budget, but I'm not pulling 10 pounds every 2 months either!

What else can I write about the lights in this place? You already know not to touch the bulbs with your fingers as the oils from your hands will imprint on the bulb and cause them to fail.

Some high end light systems run on wi-fi. You can make adjustments to your grow room with your cell phone. Pretty cool technology.

Overall, the time spent farming a four light set up is a lot like a part time job with excellent benefits.

You got this.

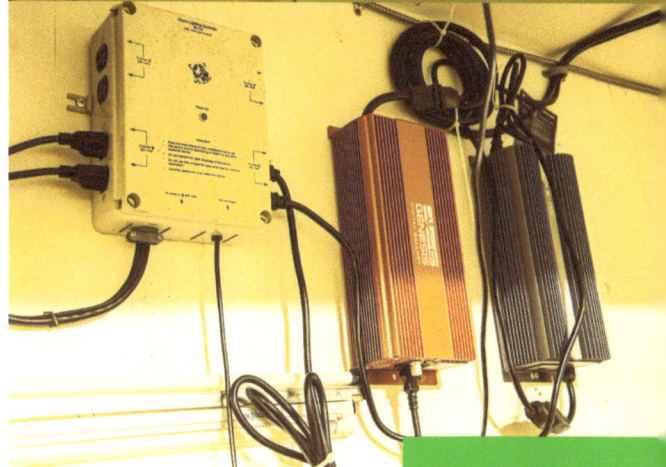

WATERING SYSTEM. This room feeds the plants using an ebb and flow spray system. It works like a traditional ebb and flow system but instead of flooding the grow tray from below, it directly sprays the Rockwool™ cubes from above. Upon saturation the cubes release the excess nutrient water solution into the tray which then drains back to the reservoir. This recirculating sprayer system uses a **water chiller** to keep the nutrient solution at the optimum temperature of 68° F. The pump is plugged into a timer and when activated, feeds the entire garden.

Note the different mediums and systems in use. Room one has the oxygen infused Aero-Flo™ system with its 3-inch baskets full of rock medium; Room two hand waters into 10 gallon pots filled with soil-less coco mix; Room three's medium is Rockwool™ cubes on top of Rockwool™ slabs while using sprayers to feed. Rockwool™ is a good medium to work with because it will release excess water once the cube has been saturated. It is because of it's ability to release water that it requires frequent feedings which is good for the plants as they get fresh oxygen and nutrients delivered to the root system multiple times throughout the day. During peak flowering, weeks four through seven, he waters this room four times per day. 4X's at 15 minutes per.

I'm not sure what watering system out of the three rooms is the best but a ten pound harvest speaks for itself.

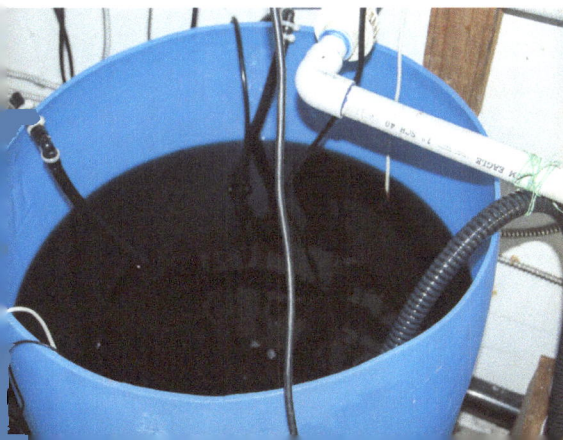

AIR ENVIRONMENT & AIR CONDITIONER. Of course, this room runs a closed system. That means no air from the outside is coming into the space. The doors are closed and the A/C unit is ON and set at 82°F. For production to stay at maximum levels, running a cooled room is critically important. For this reason, the grower has installed a second "emergency" A/C unit as a backup if the primary unit goes down. There will be no lost harvest due to an air conditioner malfunction in room three.

Room three is using a 2-ton, 24,000 BTU, mini-spilt, air conditioner unit similiar to the one in room two. This is the smallest system that works well with a four light, 1000 watt, HPS set-up. Also notice the cardboard attachment in the picture below. The grower added this cardboard to help direct the cold air above the canopy. He realized the cold air blasting from the A/C unit was too cold for the girls! He had to make the adjustment so that cold air was

not harming the plants. As the grower in your space you'll have to make judgement calls in your room so that production stays on target. If you think something is not good for the plants... you are probably correct in making the beneficial adjustments.

This picture is of the second stand-by A/C unit. It only kicks ON if the primary unit has a malfunction and the room begins to heat up.

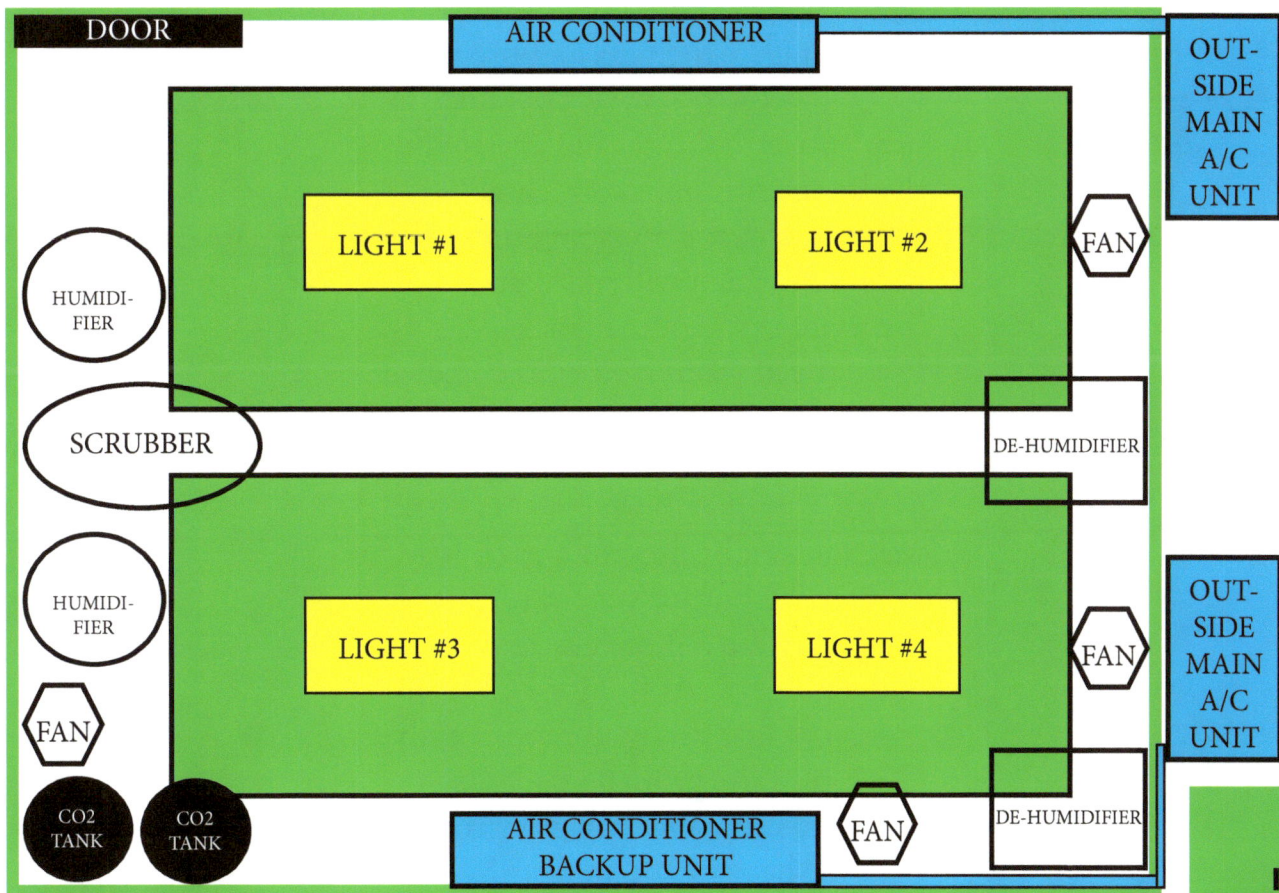

CO2 System. Room three uses CO2 tanks instead of a CO2 generator / burner. He keeps the CO2 levels in the room between 1200 PPM and 1300 PPM during the intense growing periods of weeks one through six. He drops the CO2 level down to between 800 PPM and 900 PPM for the last two weeks of flowering and ripening.

Let's go over the differences between the two CO2 systems. CO2 generators convert O2 into CO2. They do this conversion with fire. CO2 generators have burners inside them that light up with a flame and "convert" the oxygen into carbon dioxide.

CO2 tanks differ in that there is no flame, no fire is used to convert the gas but rather the tanks release pure CO2 into the room. CO2 tanks do not "convert" the gas but merely fill the room with pure condensed CO2. This can create a pressure build up in the room which will push air out of the room through the cracks.

I've noticed growers in cold climates seem to use CO2 generators with burners and growers in warm climates use CO2 tanks. Both systems work well and can be used in any climate. I prefer the burner system because I think they are a more efficient system as the unit coverts O2 into CO2 instead of just filling the room with CO2, plus the propane tanks are lighter to carry into my grow room than the heavy CO2 tanks.

OTHER AIR ENVIROMENTAL TOOLS. Remember, running a closed system would not be possible without the use of a computer and sensors. The grow industry has manufactured CO2 systems for over 40 years, but it was not until the technology had advanced enough with computers and sensors that the average farmer was able to use and benefit from the closed system. The CO2 controller computer, with its sensors, constantly measures the CO2 levels in the room. When the computer senses that the CO2 levels have dropped below the programed level of 1300 parts per million, it activates the CO2 generator (or tank). Then, once the appropriate levels have been achieved, it de-activates the generator (or tank).

The computer system is working all day doing this. Without it, the grower would have to "live" in the grow room and turn the generators and fans ON and OFF manually. This, in reality, is not possible. Lucky for us, there are several CO2 controller systems on the market from which to choose. I actually do not have the Sentinel™ CPPM-1/CO2 PPM Controller like both the growers in room two and three. I have a small, inexpensive, plug-in unit manufacured by Hydrofarm™, the Autopilot™ CO2 controller, and it does an excellent job for me. The Sentinel™ unit does a great job, but you'll pay more money for it. In the end, it does not matter what brand you buy, as long as it works turning the CO2 unit ON and OFF at your desired levels.

MORE AIR ENVIROMENTAL TOOLS. Room three is on the next level. If you are trying to take your room there, then you'll want to employ a humidifier(s) and a de-humidifier(s). You need these tools if you want control of the humidity levels in the grow space. Grower three keeps his room at a higher humidity level than most growers are accustomed to running their grows. I think this is a big reason his room is producing two to three times more flowers then the other rooms.

He keeps the humidity levels in this room an average of 70% during the first 6 weeks of bloom. Most growers are too scared of mold and powdery mildew to run their rooms at these higher levels of humidity, but if you want your plants to grow like they do in **Jamaica** then you

have to turn your room into **Jamaica.** There is a system, a better grow environment for the plants that uses a temperature and humidity range that growers try to keep their rooms in to maximize production. It is the V.P.D. and I'll explain that on page 43.

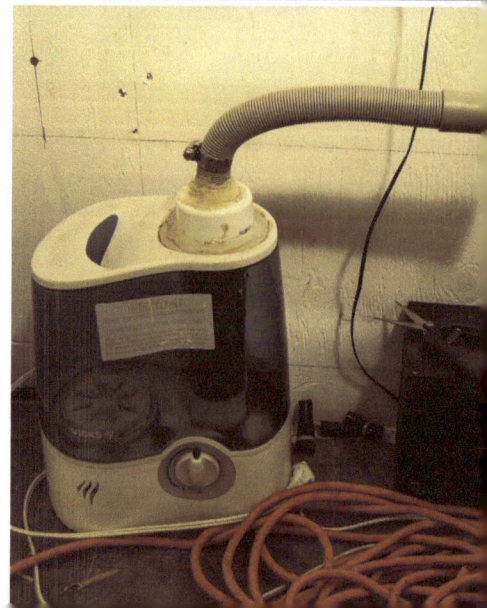

CUSTOM SCRUBBER FAN SYSTEM. Grower three's spot is in the city and you better believe he uses a scrubber to keep the smells down. Scrubbers are a necessity unless you don't care about stinking up the whole neighborhood. Most smart growers use them but this grower has gone genius on the subject. Notice the metal vent with holes in the picture below. He has turned his scrubber into an air movement system, a giant room fan! It is actually really cool, as it performs two tasks; one is cleaning the skunk air odor and the other task is blowing air onto the canopy.

Air Holes

End Cap

VPD. Vapor Pressure Differential (or Deficit) is a mathematical equation that helps growers reach and maintain the optimal humidity and temperature levels for maximum growth and yield rates. VPD is the difference between the saturated vapor pressure (VPsat) minus (-) the air vapor pressure (VPair). This equation helps determine the amount of humidity pressure the air is putting on the plants. If the **humidity** in the grow is **low**, there will be little atmospheric pressure pushing back on the plant leaves and they will start to **uncontrollably sweat/transpire** and this is not good for growing. Reversely, if the **humidity** in the grow room is too **high**, there will be heavy atmospheric pressure pushing so hard on the leaves so that they will **not be able to sweat/transpire.** This situation is also bad for growing. It is the atmospheric pressure that does the pushing and pulling on the leaves. The atmospheric pressure of the grow room is the combination of the temperature and humidity. It's easy if you follow the graph but if you want to make it more complicated, just do the math...

$$VPsat = \frac{610.7 * 10((7.5*(Leaf\ Temp))/(237.3+(Leaf\ Temp))}{1000}$$

You take that VPsat number and minus (-) it to this VPair equation:

$$VPair = \frac{610.7 * 10((7.5*(Air\ Temp))/(237.3+(Air\ Temp))}{1000} \times ((Room\ Humidity)/100)$$

If you follow these equations, you will come up with a number. When trying to follow the VPD you'll want keep the temperature and humidity in your room at levels that when plugged into this equation will equal (1.3 - 1.8). The sweet spot is plus or minus 0.5 in either direction. This means the grower wants to keep the product of this mathematical equation between the numbers of 0.7 to 2.1 to be in the preferred zone for plant health.

It is good to know the complicated math and how we've come to this knowledge but in the end, as a grower, you want to remember these basic rules...

A low humidity will give you a high VPD number (above 2.0) which is not good.
A very high humidity will give you a very low VPD (below 0.5) which is not good.
Grow in the sweet spot!

VPD reference table (values in kPa). RH across the top (%), Temp down the left side (°F).

Temp (°F)	100	98	96	94	92	90	88	86	84	82	80	78	76	74	72	70	68	66	64	62	60	58	56	54	52	50	48	46	44	42	40	38	36	34	32	30	28	26	24	22	20	18	16	14	12	10	8	6	4	2	0
50	0	0	0	0.1	0.1	0.1	0.1	0.2	0.2	0.2	0.2	0.3	0.3	0.3	0.3	0.4	0.4	0.4	0.4	0.5	0.5	0.5	0.5	0.6	0.6	0.6	0.6	0.7	0.7	0.7	0.7	0.8	0.8	0.8	0.8	0.9	0.9	0.9	0.9	1	1	1	1.1	1.1	1.1	1.1	1.1	1.2	1.2	1.2	1.2
52	0	0	0.1	0.1	0.1	0.1	0.2	0.2	0.2	0.2	0.3	0.3	0.3	0.4	0.4	0.4	0.4	0.5	0.5	0.5	0.6	0.6	0.6	0.7	0.7	0.7	0.7	0.8	0.8	0.8	0.9	0.9	0.9	0.9	1	1	1	1	1.1	1.1	1.1	1.2	1.2	1.2	1.3	1.3	1.3
54	0	0	0.1	0.1	0.1	0.1	0.2	0.2	0.2	0.3	0.3	0.3	0.3	0.4	0.4	0.4	0.5	0.5	0.5	0.6	0.6	0.6	0.6	0.7	0.7	0.7	0.8	0.8	0.8	0.8	0.9	0.9	0.9	1	1	1	1.1	1.1	1.1	1.1	1.2	1.2	1.3	1.3	1.3	1.3	1.4	1.4
55	0	0	0.1	0.1	0.1	0.1	0.2	0.2	0.2	0.3	0.3	0.3	0.4	0.4	0.4	0.4	0.5	0.5	0.5	0.6	0.6	0.6	0.7	0.7	0.7	0.7	0.8	0.8	0.8	0.9	0.9	0.9	1	1	1	1.1	1.1	1.1	1.2	1.2	1.3	1.3	1.3	1.3	1.4	1.4	1.4	1.5
57	0	0	0.1	0.1	0.1	0.1	0.2	0.2	0.2	0.3	0.3	0.4	0.4	0.4	0.5	0.5	0.5	0.6	0.6	0.6	0.7	0.7	0.7	0.8	0.8	0.8	0.9	0.9	0.9	1	1	1.1	1.1	1.1	1.2	1.2	1.3	1.3	1.3	1.4	1.4	1.5	1.5	1.5	1.6	1.6
59	0	0	0.1	0.1	0.1	0.1	0.3	0.3	0.3	0.3	0.4	0.4	0.5	0.5	0.5	0.6	0.6	0.6	0.7	0.7	0.8	0.8	0.9	0.9	1	1	1.1	1.1	1.1	1.2	1.2	1.3	1.3	1.5	1.5	1.6	1.6	1.6	1.7	1.7
61	0	0	0.1	0.1	0.1	0.2	0.2	0.3	0.3	0.4	0.4	0.5	0.5	0.5	0.6	0.6	0.7	0.7	0.8	0.8	0.9	0.9	1	1	1.1	1.1	1.2	1.2	1.3	1.3	1.4	1.4	1.5	1.5	1.6	1.6	1.7	1.7	1.7	1.8	1.8
63	0	0	0.1	0.2	0.2	0.2	0.3	0.3	0.4	0.4	0.5	0.5	0.5	0.6	0.6	0.7	0.7	0.8	0.8	0.9	0.9	1	1	1.1	1.1	1.2	1.2	1.3	1.3	1.4	1.4	1.5	1.5	1.6	1.6	1.7	1.7	1.8	1.9	1.9	1.9
64	0	0	0.1	0.2	0.2	0.2	0.3	0.3	0.4	0.4	0.5	0.5	0.6	0.6	0.7	0.7	0.7	0.8	0.9	0.9	1	1	1.1	1.1	1.2	1.2	1.3	1.3	1.4	1.4	1.5	1.6	1.6	1.6	1.7	1.7	1.8	1.9	1.9	2	2.1
66	0	0.1	0.1	0.2	0.2	0.3	0.3	0.4	0.4	0.5	0.5	0.6	0.6	0.7	0.7	0.7	0.8	0.8	0.9	0.9	1	1	1.1	1.2	1.2	1.3	1.3	1.4	1.5	1.5	1.6	1.6	1.7	1.8	1.9	1.9	2	2.1	2.1	2.2
68	0	0.1	0.1	0.2	0.2	0.3	0.3	0.4	0.5	0.5	0.6	0.6	0.7	0.7	0.8	0.9	0.9	1	1	1.1	1.1	1.2	1.3	1.3	1.4	1.4	1.5	1.5	1.6	1.6	1.7	1.7	1.8	1.8	1.9	1.9	2	2	2.1	2.1	2.2	2.2	2.3
70	0	0	0.1	0.1	0.2	0.3	0.3	0.4	0.4	0.5	0.6	0.6	0.7	0.7	0.8	0.8	0.9	0.9	1	1.1	1.1	1.2	1.2	1.3	1.4	1.4	1.5	1.6	1.6	1.7	1.7	1.8	1.9	1.9	2	2	2.1	2.1	2.2	2.2	2.3	2.4	2.4	2.5
72	0	0.1	0.1	0.2	0.3	0.3	0.4	0.5	0.5	0.6	0.6	0.7	0.7	0.8	0.9	0.9	1	1.1	1.1	1.2	1.2	1.3	1.4	1.4	1.5	1.6	1.6	1.7	1.7	1.8	1.9	1.9	2	2.1	2.1	2.2	2.2	2.3	2.4	2.4	2.5	2.6	2.6
73	0	0.1	0.1	0.2	0.3	0.3	0.4	0.5	0.5	0.6	0.6	0.7	0.7	0.8	0.9	0.9	1	1.1	1.1	1.2	1.3	1.3	1.4	1.5	1.5	1.6	1.6	1.7	1.8	1.9	2	2.1	2.1	2.2	2.2	2.3	2.3	2.4	2.5	2.5	2.6	2.6	2.7	2.7	2.8
75	0	0.1	0.1	0.2	0.3	0.4	0.4	0.5	0.6	0.6	0.7	0.7	0.8	0.9	0.9	1	1.1	1.1	1.2	1.3	1.3	1.4	1.4	1.5	1.6	1.7	1.7	1.8	1.9	2	2.1	2.1	2.2	2.3	2.3	2.4	2.5	2.6	2.6	2.7	2.8	2.8	2.9
77	0	0.1	0.1	0.2	0.3	0.4	0.5	0.6	0.6	0.7	0.8	0.8	0.9	0.9	1	1.1	1.1	1.3	1.3	1.4	1.5	1.6	1.6	1.7	1.8	1.9	1.9	2	2.1	2.1	2.2	2.3	2.3	2.4	2.5	2.6	2.6	2.7	2.8	2.8	2.9	3	3.1	3.1
79	0	0.1	0.1	0.2	0.3	0.4	0.5	0.5	0.6	0.7	0.7	0.8	0.9	0.9	1	1.1	1.2	1.3	1.3	1.4	1.5	1.6	1.7	1.8	1.9	1.9	2	2.1	2.2	2.3	2.3	2.4	2.5	2.6	2.7	2.7	2.8	2.9	2.9	3	3.1	3.1	3.2	3.3	3.3
81	0	0.1	0.1	0.2	0.3	0.4	0.5	0.6	0.6	0.7	0.7	0.8	0.9	1	1.1	1.1	1.2	1.3	1.4	1.5	1.6	1.6	1.7	1.8	1.9	2	2.1	2.1	2.2	2.3	2.4	2.5	2.6	2.7	2.7	2.9	3	3.1	3.1	3.3	3.3	3.4	3.5
82	0	0.1	0.2	0.2	0.3	0.4	0.5	0.5	0.6	0.7	0.8	0.8	0.9	1	1.1	1.2	1.3	1.4	1.4	1.5	1.6	1.7	1.7	1.8	1.9	2	2	2.1	2.2	2.3	2.3	2.4	2.5	2.6	2.6	2.7	2.8	2.9	2.9	3.1	3.2	3.3	3.3	3.4	3.5	3.5	3.6	3.7
84	0	0.1	0.2	0.2	0.3	0.4	0.5	0.6	0.7	0.8	0.8	0.9	1	1.1	1.1	1.3	1.4	1.4	1.5	1.6	1.7	1.7	1.8	1.9	2	2.1	2.1	2.2	2.3	2.4	2.5	2.6	2.6	2.7	2.8	2.9	3	3.1	3.2	3.3	3.3	3.4	3.5	3.6	3.7	3.7	3.8	3.9	4
86	0	0.1	0.2	0.3	0.4	0.5	0.6	0.7	0.7	0.8	0.9	1	1.1	1.2	1.3	1.4	1.5	1.6	1.7	1.8	1.9	2	2.1	2.2	2.3	2.4	2.5	2.6	2.7	2.8	2.9	3	3.1	3.2	3.3	3.4	3.5	3.6	3.7	3.8	3.9	4	4.2
88	0	0.1	0.2	0.3	0.4	0.5	0.6	0.7	0.8	0.9	1	1.1	1.2	1.3	1.4	1.5	1.6	1.7	1.8	1.9	2	2.1	2.1	2.2	2.3	2.4	2.5	2.6	2.7	2.8	2.9	3	3.1	3.2	3.3	3.4	3.5	3.6	3.7	3.7	3.8	3.9	4	4.1	4.2	4.3	4.4	4.5
90	0	0.1	0.2	0.3	0.4	0.5	0.6	0.7	0.8	0.9	1.1	1.2	1.3	1.4	1.5	1.6	1.7	1.8	1.9	2	2.1	2.2	2.3	2.4	2.5	2.6	2.7	2.8	2.9	3	3.1	3.2	3.3	3.4	3.5	3.6	3.7	3.8	4	4.1	4.2	4.3	4.4	4.5	4.6	4.7
91	0	0.1	0.2	0.3	0.4	0.5	0.6	0.7	0.8	0.9	1.1	1.1	1.2	1.3	1.4	1.5	1.6	1.7	1.8	1.9	2	2.1	2.2	2.3	2.4	2.5	2.6	2.7	2.8	2.9	3	3.1	3.2	3.3	3.4	3.5	3.6	3.7	3.8	3.9	4	4.1	4.2	4.3	4.4	4.8	4.9	5
93	0	0.1	0.2	0.3	0.4	0.5	0.6	0.7	0.8	1	1.1	1.3	1.4	1.5	1.6	1.7	1.8	1.9	2	2.2	2.3	2.4	2.5	2.6	2.7	2.8	3	3.1	3.2	3.3	3.4	3.5	3.7	3.8	3.9	4	4.1	4.3	4.4	4.5	4.6	4.7	5	5.1	5.2	5.3
95	0	0.1	0.2	0.4	0.5	0.6	0.7	0.8	0.9	1	1.1	1.2	1.3	1.4	1.6	1.7	1.8	1.9	2	2.1	2.2	2.3	2.5	2.6	2.7	2.8	2.9	3	3.1	3.2	3.3	3.5	3.6	3.7	3.8	4	4.1	4.2	4.3	4.4	4.6	4.7	4.9	5	5.1	5.2	5.4	5.5	5.6
97	0	0.1	0.2	0.4	0.5	0.6	0.7	0.8	0.9	1	1.1	1.2	1.3	1.5	1.6	1.8	1.9	2	2.1	2.2	2.4	2.5	2.6	2.7	2.8	2.9	3.1	3.2	3.3	3.4	3.5	3.7	3.8	3.9	4	4.2	4.4	4.5	4.6	4.7	4.9	5.1	5.2	5.4	5.5	5.7	5.8	5.9
99	0	0.1	0.2	0.4	0.5	0.6	0.7	0.9	1	1.1	1.2	1.3	1.4	1.5	1.6	1.7	1.9	1.9	2.1	2.2	2.4	2.5	2.6	2.7	2.9	3	3.1	3.2	3.3	3.4	3.6	3.7	3.9	4	4.1	4.2	4.4	4.6	4.7	4.9	5.1	5.2	5.4	5.5	5.6	5.7	5.8	6	6.1	6.2
100	0	0.1	0.3	0.4	0.5	0.7	0.9	1.1	1.2	1.3	1.4	1.6	1.7	1.8	2	2.1	2.2	2.4	2.5	2.6	2.8	2.9	3	3.2	3.3	3.5	3.7	3.9	4.1	4.2	4.3	4.4	4.6	4.7	4.9	5	5.1	5.3	5.4	5.5	5.8	5.9	6	6.2	6.3	6.4	6.6
102	0	0.1	0.3	0.4	0.6	0.8	1	1.1	1.2	1.4	1.5	1.7	1.8	1.9	2.1	2.2	2.4	2.5	2.6	2.8	2.9	3.2	3.3	3.5	3.6	3.7	3.9	4.2	4.3	4.4	4.6	4.7	4.9	5.1	5.3	5.4	5.5	5.7	5.8	6	6.1	6.4	6.5	6.7	6.8	6.9
104	0	0.1	0.3	0.6	0.7	0.9	1.2	1.3	1.5	1.6	1.8	1.9	2.1	2.3	2.4	2.6	2.8	2.9	3.1	3.2	3.4	3.6	3.7	3.8	3.9	4.1	4.3	4.4	4.6	4.8	4.9	5.1	5.3	5.4	5.6	5.7	5.8	6	6.1	6.3	6.4	6.6	6.8	6.9	7	7.2	7.3
106	0	0.2	0.3	0.5	0.6	0.8	0.9	1.1	1.2	1.4	1.5	1.7	1.8	2	2.2	2.3	2.5	2.6	2.8	2.9	3.1	3.2	3.4	3.5	3.7	3.9	4	4.2	4.3	4.5	4.6	4.8	4.9	5.1	5.2	5.4	5.5	5.7	5.9	6	6.2	6.3	6.5	6.6	6.8	6.9	7.1	7.2	7.4	7.6	7.7
108	0	0.2	0.4	0.5	0.7	0.9	1	1.1	1.2	1.4	1.5	1.6	1.8	1.9	2.1	2.3	2.4	2.6	2.8	2.9	3.1	3.2	3.4	3.6	3.7	3.9	4.1	4.2	4.4	4.5	4.7	4.9	5.2	5.4	5.5	5.7	5.8	6	6.2	6.5	6.7	7.1	7.3	7.5	7.6	7.8	8.1
109	0	0.2	0.3	0.5	0.7	0.9	1	1.1	1.2	1.4	1.5	1.7	1.9	2.1	2.2	2.4	2.6	2.7	2.9	3.1	3.3	3.4	3.6	3.8	3.9	4.1	4.3	4.4	4.6	4.8	5.1	5.5	5.6	5.8	6	6.2	6.3	6.5	6.7	6.8	7	7.2	7.4	7.5	7.7	7.9	8	8.2	8.4	8.6
111	0	0.2	0.4	0.5	0.7	0.9	1.1	1.3	1.4	1.6	1.8	2	2.2	2.3	2.5	2.7	2.9	3.1	3.2	3.4	3.6	3.8	4	4.1	4.3	4.5	4.7	4.9	5.2	5.4	5.6	5.8	5.9	6.1	6.3	6.5	6.7	6.8	7	7.2	7.4	7.6	7.8	7.9	8.1	8.3	8.5	8.7
113	0	0.2	0.4	0.6	0.8	0.9	1.1	1.3	1.5	1.7	1.9	2.1	2.3	2.5	2.7	2.8	3	3.2	3.4	3.6	3.8	4	4.2	4.4	4.6	4.7	4.9	5.1	5.3	5.5	5.7	5.9	6.1	6.3	6.5	6.6	6.8	7	7.2	7.4	7.6	7.8	8	8.2	8.4	8.7	8.9	9.1	9.3	9.5

VPD Graph Courtesy of Pulse Labs.
Go to Pulsegrow.com and check out the BEST environmental monitor available today and use the product code: mrgreen$50
for $50 off your next order.

Keep your humidity and temperatures in the green-golden range of this grow graph. At first you don't need to fully understand it, just do it, your comprehension will come.

• Follow the graph and control your temperature and humidity so that you are growing in the sweet spot.

• Mastery of this concept gives the grower more information on the true health of the plants. With this critical information you can control your closed system to achieve the best results possible.

• Use this graph when trying to achieve the perfect balance between the humidity and temperature in your grow room.

THE HARD PRUNE.

If you want your garden canapy to look like this...

YOU SHOULD CONSIDER HARD PRUNING LIKE THIS...

Grower number three grows for the flowers (not for wax nor edibles) and he hard prunes his crop, weeks 1 and 3 of budding. If you are not already doing this, then it is going to freak you out. So, I recommend experimenting on one plant first to see if you like the results. This technique is best for strong strains that do not stress easily. If your strain hermaphrodites when stressed, then you may want to hold back on this trick until you get some stable genetics. A HARD PRUNE is when the grower removes ALL the large and medium sized fan leaves and trims off ALL the lower third (1/3) branches. It looks strange at first but the leaves all grow back within a week, so don't be scared.

After the prune job, he weaves the remaining branches in and out of the canopy trellis which soon grows into a table top of flower buds.

Trimming the plants back like this does many things to them and some growers brag that it is this technique that gives them the weight advantage over other growers. I'm not sure about the "huge" overall yield gains but I do know that this technique has its advantages. I think removing most of the leaves can push the harvest date back up to a week because the plants have to re-grow all the fan leaves and this takes time. After the first prune, while the leaves are trimmed off, the plant gets massive light on its entire body which creates many bud-set locations. After the second prune, these buds are once again fueled by direct light until some of the leaves start to grow back. This hard pruning technique obviously affects the intetrmal mechanisms of the plants, so that most of their energy is working towards producing flowers. For years, I never cut a branch nor leaf. I thought it better just to leave the plants alone. Now, I've been growing close to 30 years and I hard prune my plants like a bonsai tree. I feel like I get better control over the plants, with less bug problems, and a much easier ALL NUG harvest. Try the hard prune. It just may work for you.

NUTRIENTS.

You would think that the brand of nutrient you use to grow makes a big difference in the harvest results. This is true to some degree but if you are purchasing top tier nutrients from a "grow shop" the differences between nutrient brands becomes less and less. In my opinion, professional grade cannabis nutrients are more similar than different, and it's the products marketing that sets them apart. With that said...

Room 3 uses: Heavy 16™,Hydrozyme™, KoolBloom™ & teas.
Room 2 uses: Dutch Master™, Canna Coco™.
Room 1 uses: Botanicare™.
All of these brands are professional grade, top tier cannabis nutrients and usually produce excellent crop results. But grower three will argue that his choice of nutrients are the best and he has 10 pounds of chronic to prove it. It's hard to argue with that kind of result.

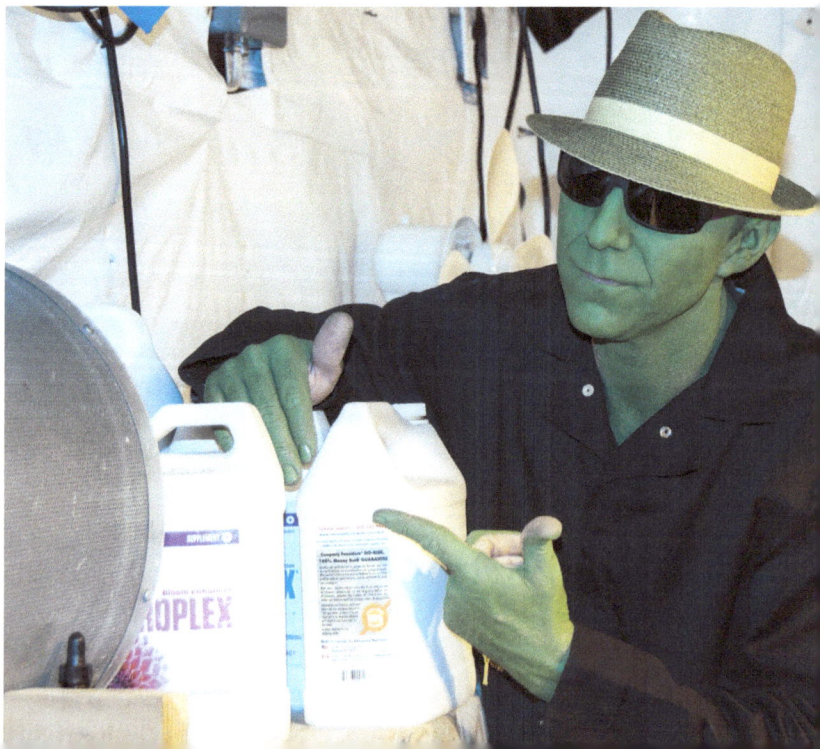

TEA TIME: An introduction to root microbiology. I think the best techniques, that get the best results, are the ones that listen to the plants. When growing indoors, in a hydroponic system set-up, I feel the plants are yelling out for something. They need something more than mere nutrients to thrive. They are in need of some microbiology! Think of plant microbiology as probiotics for plants. When growers talk about teas they are not talking about chamomile. Plant teas are mixes of beneficial micro bacteria and micro fungi that are "brewed" together in an oxygen bath of organic soup. These teas are intended for the plants root system and should be delivered independent of the nutrient feed. The high salts necessary to stabilize bottled nutrients actually damage and kill the delicate micro-biology of plants. Likewise, the delicate micro bacteria can also get harmed while being pumped through spray nozzles. So, if you want the micro-biology to actually get down to the root system and benefit the plants, then you should consider first, watering the teas by hand and second, watering them at different times, separate from nutrient feeds. The product he uses with great success is a five gallon pre-made tea from Xtreme Gardening™. Great White™ Mycorrhizae works good too.

PEAK HARVEST TECHNIQUES. The trick is knowing when to say when. Cannabis plants generally finish flowering between eight and twelve weeks from the first day of the bloom stage (12 hours of light). Pure indicas can finish in eight weeks, while strong sativas can take up to twelve weeks to ripen. Mark the calendar on the first day of flowering cycle so that you'll know how long the plants will take to finish budding. Once you've worked with a strain for a few cycles, you'll be able to better predict the harvest day more accurately. You'll want to start checking the crystals on the plant about one week before the scheduled harvest date. By this time, most of the pistils (long white hairs) on the bud will have changed colors from white to amber, brown or red thus signaling to the grower that peak harvest time is near. When searching for the peak THC ripeness in the crystals, look for the crystals to be stacked, erect and swollen. The crystals will be filled with a clear/white creamy looking substance and will have a head/cap

on its top. This is as good as its going to get and if 2/3rds of the plant looks like this, then it is time to harvest. You may find that some of the buds on a plant are still looking unripe, in which case you can ask yourself "do all the apples on the apple tree come in ripe on the same day"? The answer is "NO" and this is the same with the cannabis plant too. The grower can always leave the bottom unripe buds on the plant for some more time so that they can ripen on the vine.

Pre-harvest crystals -Visible formed crystals but not swollen, more clearish.

Peak-harvest crystals -swollen, tall, head on top, more milky/white color.

Post-harvest crystals -yellow-ish in color, bent, broken, slight deflating.

When harvesting a plant, some buds may be more ripe than others and it is up to the grower to make a judgment call. Once 70 percent of the plant has hit the peak then it's time to chop her down. Some growers will harvest the ripe tops of the plants while they let the bottoms grow for a few more days to catch up before harvesting them. When it comes to a slow, relaxed harvest, hobbyists have the advantage over commercial growers because the latter is always pressured for money and time.

I was tripping out one day and realized, different harvest times can produce different head highs from the same plant. Every strain is different, thus results may vary, but I've noticed in general, pre-peak harvest buds can be a hyper-active experience, while post-peak harvest buds can be more slow and mellow of an experience. Buds harvested at the peak time perfectly blend it all together. As a grower, I invite you to experiment with this, as you may find that you prefer your flowers one way over another. I perfer my SkunkOG at peak/post-peak harvest time and my Pineapple Express slightly pre-peak harvest time. Test your strains to find out what you like best. *From Left to right...pre-peak, peak, post peak.*

MOLD PREVENTION. You might think running a closed system will raise the risk of molds, such as powdery mildew and botrytis (bud rot), ruining your garden, but we have found this not to be the case. In fact, running the room at the optimal VPD will raise the endogenous (immune) system of the plants and help them fight off these molds and other diseases. With this said, you'll want to do preventative measures to insure your crop avoids these pitfalls. It all starts in the soil! White powdery mildew actually lives in the soil. You can fight and defeat the white-spore, hyphae infection on the leaves, only to discover a new outbreak a few days later because the plant has been infected in the roots and soil.

• Start the grow with a clean grow room and good soil or hydro medium that is not infected with mold spores.
 • Try to Hard Prune the garden by removing large and meduim fan leaves on weeks one and three.
• Control the VPD in the grow room so that it is riding a little high, 1.5VPD - 1.75VPD, by lowering the humidity just a bit from say 70% to 55% and by not letting the temperature climb too high.
• Avoid high humidity and drastic temperature dips when the lights go off, as dusk is the dangerous time in regards to spores developing and molds taking hold. During growth and heavy budding try to keep the night time temperatures as close to the daytime temperatures as possible. The best evening protocol is when you can drop the humidity 10%, turn off the CO_2 and keep the night temperature within 0-5 degrees (Fahrenheit) of the daytime temperature.

Never rule out sulfur burners to aid you in battle against damaging molds and/or bugs. The sulfur gas changes the pH in your grow room so that molds and bugs can not survive. Only use the gas once a week for a few hours at dusk (lights off) time. Stop using sulfur gas once flowers start to develop. If you are still unlucky and get a little mold on your crop after harvest, you can always use an extraction method that uses extreme temperature fluctuations or toxic environments to kill all micro organisms in order to make concentrates.

Remember, some cannabis strains are more prone to molding in certain environments, so grower beware.

CONCLUSION

Look what is possible for your garden using the CLOSED SYSTEM. We went from grow room number one, producing 3.25 pounds, to grow room number two, which produced 4.5 pounds, to the third and final grow room weighing in over 10 pounds of top shelf flower.

Room one really put out some excellent flower, but lost in yields due to the ineffecient CO2 system. Everytime that room gets hot, the exhaust fans clear the air in the room. The CO2 never has a lot of time to saturate the plants and therefore the plants do not get the 25% increase in production.

Room two is running the closed system and getting that 25% increase in yield using the same genetics as room one. It all starts with the air conditioner as you cannot run a closed system without one. It does take a little more money and time setting things up, but once it's running correctly... the rewards will soon follow.

Room three is MIND BLOWING. Room three is producing 2x's room two and 3x's room one! How, with basically the same square footage grow area and 500 watts more light, does the grower in room three pull this off? Well, the genetics were different in room three but not that much different, as top shelf, grade A indoor cannabis strains all tend to throw comparible numbers. So maybe "Gorilla Glue" produces a little more weight than "Pineapple Express" but not pounds more, not enough to explain a 6 pound difference. No, that's not it. The reason room three kicks such ass is because the grower uses all of his knowledge to run the room more efficiently. When I write efficiently I'm not referring to the power bills, I'm referring to the room's environment, and more importantly, how his cannabis plants respond to this environment. Along with 500 more watts of double ended HPS light in the room, the grower focused on three things to achieve his success:

- *He has total control of the Air Environment, running a CLOSED SYSTEM within the optimum range of the Vapor Pressure Differential.*

- *He uses HARD PRUNING techniques to maximize production.*

- *He has introduced micro-biological teas to help the plants thrive.*

It looks like we are done here. I hope you take this information, go out there and supercharge your grow room! Get the quality and the yields you deserve.
Keep on growing.

- Mr. Just Dank Green

IGROWCHRONIC.COM • GYO420.COM

I GROW CHRONIC II

MASTERS OF THE CLOSED SYSTEM

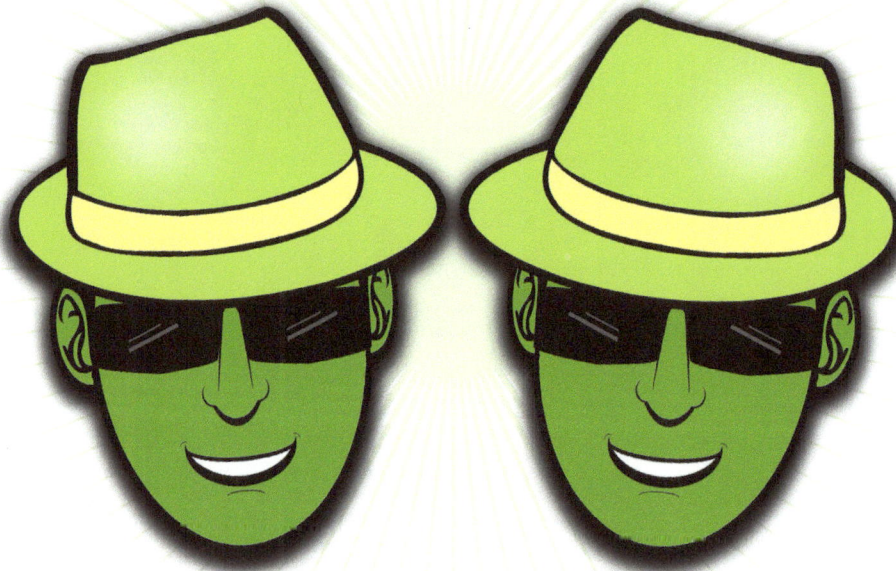

BY **MR GREEN**

IGROWCHRONIC.COM

www.ingramcontent.com/pod-product-compliance
Lightning Source LLC
Chambersburg PA
CBHW061052090426
42740CB00003B/131